GROWING THINGS

Nature study ideas for the primary school

Brian McKinlay

Illustrated by Elizabeth Honey

First published in Australia by Primary Education
Pty Ltd. in 1979.

Published in Great Britain by Studies in
Education Ltd. in 1980.

Brian McKinlay

Book Design
and Illustrations by Elizabeth Honey

Printed in Great Britain by Cheshire Typesetters Ltd.
Chester.

ISBN 090 5484 290
£2.95

Contents

Rationale

The purpose of these activities is to give the children an understanding of a number of important scientific principles, as well as to give them a chance to acquire a range of skills which will, in turn, give them confidence when undertaking simple horticultural tasks. The activities in this book may be undertaken in any order the teacher wishes, but those in the front of the book are the most simple and are, therefore, probably the best ones with which to begin.

Materials

For raising plants from seeds a great many materials and containers can be recycled and used. These will include:

glass jars of all shapes and sizes
plastic containers originally used for ice cream, margarine
egg cartons
any small wooden or metal boxes
tins used for holding coffee, tea, spices
milk cartons, paper or plastic drink cartons or containers

For holding larger, growing plants, there are a number of waste materials which can be used. In addition to terracotta or plastic pots, children can use large tins, old plastic or metal buckets, car tyres, barrels or old baskets. Also, any glass containers are useful, including old fish tanks, wide-necked jars or bottles or flat glass trays, and old kitchenware; pots, mugs, saucers or bowls.

Equipment

Little special equipment is needed for the classroom. Wooden skewers and a few knitting needles are handy for many tasks, such as planting seeds or bulbs. Some kitchen knives, spoons, and a small sprayer or atomizer are useful.

A table, bench or shelf are ideal locations for the activities. If these are not available a sunny ledge, or corner of a room or corridor, would be suitable. If possible, put plastic sheeting or old linoleum underneath pots and containers to minimize spillage onto floors or floor coverings when watering plants.

12

Soil

Soil, the basic medium of the gardener, may be either the local loam or any of the various mixtures available from nurseries. Generally, seeds do best in a light, sandy soil which will usually drain quickly after watering.

If circumstances permit, it may be possible to establish a compost heap into which vegetable wastes, leaves, grass clippings and similar materials can be placed. When these have rotted down, the soil formed can be mixed in with other potting soil.

In addition to potting soil some clean sand, small stones or fine gravel will be useful. They can be used to line containers and provide drainage. Small pieces of wood charcoal are also good for this purpose.

Dry, pulverised and relatively odourless manures can be purchased in small bags from most nurseries. Cow manure is probably the best and easiest to use, being available in finely powdered form.

Seeds and plants

All seeds and plants listed in this book can be obtained from any nursery or plant shop.

In many cases the materials used can be obtained from domestic gardens and many of the seeds are gathered from fruits and vegetables bought from the greengrocer and readily available in the kitchen.

Seeds can be shaken from the head of a plant gone to seed.

Making a start

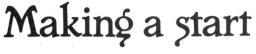

Before starting any activities using seeds, plants or bulbs, discuss with the children the way human beings rely on plants of all kinds to provide food, clothing and shelter.

Talk about gardens and discuss with the class the various reasons which lead people to grow things. Visit a local park or public garden and, if possible, look at greenhouses, ferneries, market gardens and plant shops.

Help the children to discover the whole range of industries which have to do with plants and other growing things.

Growing bean shoots or sprouts

Most children will have eaten bean shoots in Chinese food. The seeds of the Mung bean are used in producing this easily grown plant.

☐ Wash about two tablespoons of bean seeds in cold water and place them in a clean glass jar.

☐ About two or three times each day, fill the jar with cold water then drain through a strainer or muslin until the water is removed. The seeds should always be damp, but not floating in water.

☐ Keep in a light, sunny place.

☐ Within four to six days the seed will swell, then make leaf and form tiny roots. Keep up the daily routine of washing and draining.

☐ After about eight to ten days the bean shoots will have grown sufficiently to be eaten fresh or in a Chinese dish prepared by the children.

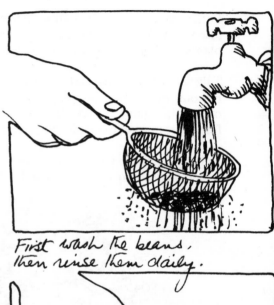

First wash the beans, then rinse them daily.

Mung beans — actual size

Bean shoots are one of the most distinctive and best known Chinese foods. Known in China for many centuries, they were a staple food for sea travellers and provided a fresh vegetable food sufficiently rich in vitamin C to prevent scurvy. This disease, caused by a vitamin deficiency, was a scourge of Western sailors but unknown to the Chinese.

Growing beans and peas

Because they germinate quickly, beans and peas are excellent subjects for indoor gardening and simple experiments.

☐ Place several beans, mixed with a handful of soil, in a large plastic bag; dampen the soil, then tie the neck of the bag. The beans will germinate inside the bag which, if left in a warm place, becomes a kind of hothouse.

☐ Place a few beans or peas between a fold of blotting paper and the side of a jar or drinking glass. Put a little water in the bottom so that the blotting paper is in contact with the water. Within a few days the beans/peas will germinate and can be seen in various stages of growth.

Beans and peas provide a good example of the ways in which the different parts of a seed grow. Children can observe the leaves, stems and roots forming and their development from the original seed.

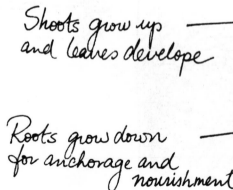

Shoots grow up
and leaves develope

Roots grow down
for anchorage and
nourishment

More activities with seeds

A number of plants can be grown in a jar. These are all basically produced in the same way as bean shoots. Such plants include:

Wheat, rice, or barley (all seeds *wholegrain* not polished)

Alfalfa or lucerne (an edible grass)

Lentils (a form of bean used as a food in southern European and middle Eastern dishes)

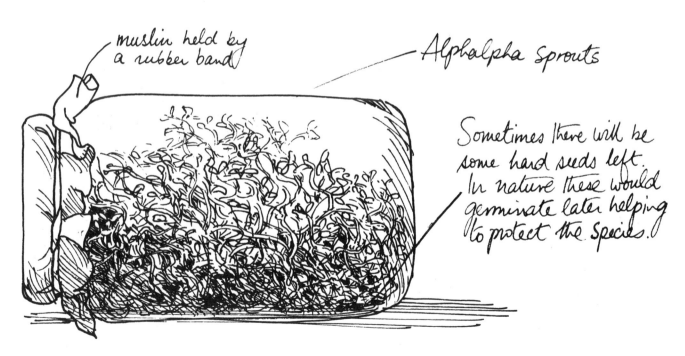

muslin held by a rubber band

Alphalpha sprouts

Sometimes there will be some hard seeds left. In nature these would germinate later helping to protect the species.

Growing seeds: wheat and rice

- ☐ Line a dish with cottonwool.

- ☐ Moisten it with water.

- ☐ Place wheat/rice upon it.

- ☐ Leave in a warm position where there is adequate light and air.

- ☐ Moisten cottonwool every few days.

- ☐ Wheat/rice will germinate within four or five days, when shoots appear.

Children can observe the way in which the wheat/rice grows; also notice and list the factors which act as limits to continued growth. For example, the different rate of growth if a dish is placed in a cold, sunless place.

Seed germination count

In this activity the children can check on the reliability of seeds and their germination rate.

Take four or five different kinds of seeds and count out equal quantities of each kind. Fairly large seeds are easier to handle and for this reason seeds of *sweet corn, sunflowers, wallflowers, sweet peas,* or any of the grains, are best.

Sow the seed either onto saucers lined with cottonwool or into small containers filled with potting soil.

The children should make a chart showing the time of planting and mark the chart each day to show the number of seeds which have germinated from day to day; also the speed of germination and the failure rate.

SEED GERMINATION CHART
All planted on Monday 2ND March

Date (20 seeds of each planted)	Number germinated				
	Sweet Corn	Sunflowers	Wall-flowers	Sweet peas	barley
Monday 9th	1	–	–	–	–
Tuesday 10th	3	–	1	–	–
Wednesday 11th	4	1	3	1	2
Thursday 12th	6	7	4	2	9

sweet corn

sunflowers

wall-flowers

sweet peas

barley

Beans and peas: some simple experiments

Seeds respond differently to varied situations. To show this principle take a number of similar containers, fill each with soil, plant a few beans or peas in each container making the following variations in their conditions:

☐ Give the first container no water at all.

☐ In the second container plant the seeds at a greater depth than in the others.

☐ Place the third container in a dark place; a cupboard would be ideal.

☐ The fourth container can be placed in a refrigerator.

☐ Place the fifth container in a warm and sunny area.

When the seeds begin to germinate, have the children make a chart showing the different times of germination; they can also measure and compare the different rates of growth.

Ask the children to draw conclusions from their experiments and have them deduce the ideal conditions for plant growth. They may devise more ways of testing the basic requirements for growth.

BEAN GROWTH CHART

	'No water' ① beans	'Deeply ② planted beans	'No light' beans ③	'refrigerator' ④ beans	'Window sill' ⑤ beans
Date planted					
Shoots appear					
Leaves appear					
How tall - 1 week					
2 weeks					
3 weeks					
4 weeks					
5 weeks					

Plants without roots

Discuss with the children the special class of plants called *moulds*. Their remarkable ability to grow under all sorts of conditions is worthy of note. Draw attention to their similarity to the fungi—mushrooms, toadstools.

Place some damp, stale bread on a dish. Leave it for several days and observe the way mould grows. Once the mould is well formed, place the bread in a jar and continue to observe the way the mould develops.

The Mould Family

Members of this family are often useful to mankind. Some moulds are added to cheese to give it special flavour and colour; for example blue cheese.

Other moulds give us penicillin, which is the most widely used antibiotic.

Yeast, an essential in bread making, is of this family too. Make a dough and add yeast to it. Instead of cooking the dough, allow the yeast to continue the fermentation process and note the effects on the dough.

"Mouldy" cheese

Baker's yeast is compressed and must be fresh

Granulated yeast keeps in an airtight container

Brewer's yeast is in fine powder form.

Plants without soil

The children will begin to understand that it is possible to grow plants without soil.

Some simple experiments can now be undertaken growing plants with large roots in water. Plants with tap roots, bulbs or tubers, make the best subjects for these experiments.

Take the tops of carrots, parsnips, beetroots and, leaving a little of the root attached, sit them, with their leaves still on, in a flat, shallow dish of water.

To help balance the plants and keep them firm, place some stones in the water too.

Plants in water

Potatoes and sweet potatoes can both be grown in water. They can be placed in a bowl or wide-necked jar and wedged into place using toothpicks or stones to ensure that the base is in the water.

Sweet potatoes grow best in a warm, sunny place and, once growing, will make a very lush indoor plant.

Carrot-fern upside down plant

Using a carrot with fronds attached, the children can grow an attractive plant and carry out an interesting experiment at the same time.

The top half of the carrot, with fronds, is suspended by string or wire. A small hollow is made in the stump of the carrot and kept filled with water.

The carrot will grow quite a luxuriant growth of leaves and make an attractive indoor plant.

Bulbs in water

Onions, daffodils, jonquils, or the smaller bulbs like shallots or freesias, can all be grown in water.

Wedge the bulb into the top of a jar or bottle with the lower half of the bulb touching the water. In a short time small roots will form and extend into the water filled container. Leaves, and sometimes flowers, will grow on the bulb.

As the plant draws only water from the container, it must draw its nutrients from material stored inside the bulb.

An interesting experiment can be carried out by replacing the water with a solution containing certain nutrients. Dissolve some sugar and starch in boiling water and, when cold, pour this mixture into the bottle from which the water has been drained. The effects on the plant can be observed.

The children can carry out a controlled experiment by giving one plant a nutrient-enriched mixture, whilst another has only water.

True water plants

The children will know that there are many plants which live all the time in water, such as pond weeds and seaweeds.

Use a small fish tank, or a large wide-mouthed jar, to grow a collection of common pond weeds; most of which grow simply floating in the water or rooted in the mud.

Ask the children to research the various products made from seaweed.

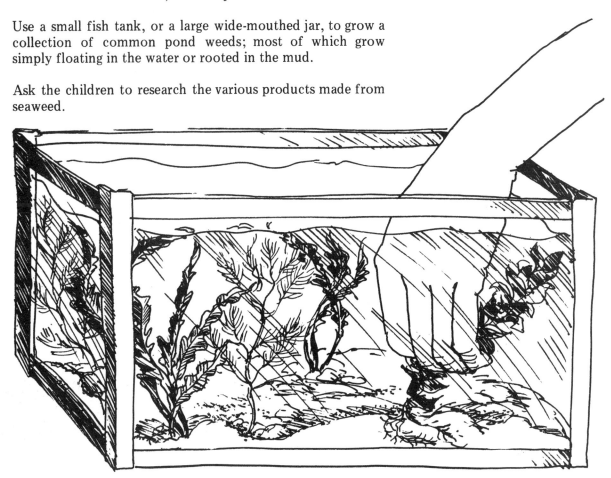

Plants and water

Simple experiments will demonstrate to the children the way plants transmit water.

☐ Take a potato, a stick of celery, some food dye, water, and two bowls.

☐ Place water in both bowls, with dye in one of them.

☐ Divide the celery, leaving the top of the stick joined.

☐ Place one part in the clear water, the other in the dyed water. Allow time for the fluids to move up through the celery.

The same experiment may be tried with two pieces of potato. One piece placed in dyed water, the other in clear water.

clean water

Coloured water

Plants and sunlight

In this activity the children take a number of small pots and place in them easily grown plants; geranium, beans, or carrots.

☐ Cover one plant with a box which has a small square cut in one side. Put the box where it will not be moved. With the passage of several days the children will notice that the plant will turn towards the light.

☐ Take another plant, one with fairly large leaves, and cover one leaf with tissue paper. Put tape over the paper. Handle the leaf gently so as not to bruise it. In a week take the tissue paper away and the children will see that the leaf, deprived of sunlight, has begun to change colour and lose its green appearance.

☐ Place a potplant inside a box and leave it, without light, for about a week. Ask the children to observe the changes which have occurred.

leaf covered with paper

Plant experiment in progress
Do not disturb!

Plants in a maze

Take a healthy bean or pea plant growing in a pot. Place an obstruction in its way as it grows. In time the plant, unable to grow upright, will grow around the obstruction.

Ask the children to observe the way plants and trees overcome natural obstacles in the world around them.

The strength of seeds

For this activity the children fill a small jar as tightly as possible with bean or pea seeds. Add a quantity of water and screw the lid on tightly.

As the seeds absorb the water they will break the jar or force off its lid.

The same experiment can be tried with a foil bag or a plastic container.

This jar was packed full of pea seeds on Wednesday (20th May) at 11 am.

The strength of plants

Take a large plastic ice-cream container, fill it with soil and plant about a dozen beans. Pour over the top of the soil a mixture of liquid plaster of Paris. Cover the soil and allow the plaster to set.

Each day pour water on the plaster and, after some days, when the seeds germinate they will begin to break the plaster from below.

The children could experiment to find what thickness of plaster is needed to prevent the seeds growing. They can also experiment to see whether it is better to plant the seeds close together or evenly distribute them over the surface of the container.

Have them watch to see which groups of seeds push through first.

Tree roots can break up a concrete path

Plaster of Paris

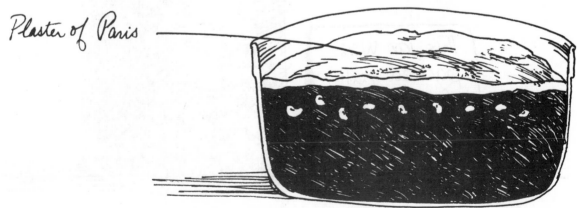

Unusual plants to grow indoors: Pineapples from pineapple tops

Place some pineapple tops on a saucer. Each top should have a small portion of the pineapple husk attached.

Keep the stump of the pineapple moist and, when rootlets begin, plant the pineapple into a small pot containing soil.

Growing avocados from seed

Keep the large seed from inside an avocado. Place it over a jar of water so that the lower part of the seed is touching the water.

After some weeks the seed will begin to sprout, and a central shoot will grow upwards. When this has happened, plant the seed and its shoot into a fairly large pot, and keep it as an indoor plant.

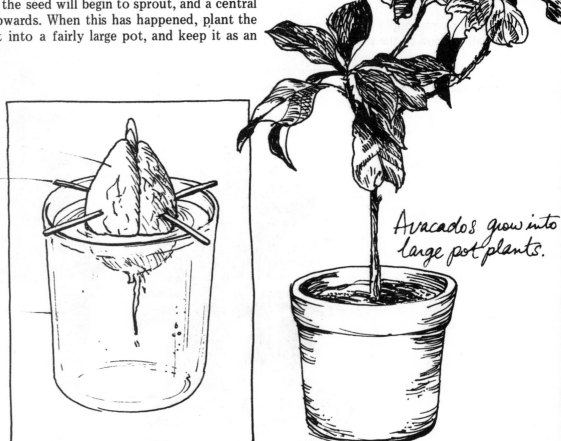

Avacado seed'

toothpick

glass of water'

Avacados grow into large pot plants.

Growing palms from seeds

Palms can be grown from the seeds to be found inside packets of dates commonly bought in supermarkets. The seeds should be planted in a pot and kept in a warm place.

Date palms grow best in a warm situation and can be kept in any heated room or near a heater of any sort. To retain heat, the pot can be kept covered in a plastic bag, which will increase the temperature of the soil and stop draughts.

Pumpkins and watermelons

These two familiar plants can be grown in containers.

Ask the children to collect seeds from inside a pumpkin or melon. Sow the seeds, either in a group in a shallow container lightly covering them with soil, or group them individually in small containers; empty drink cartons are ideal.

When established the small plants can be planted in a sunny place outdoors.

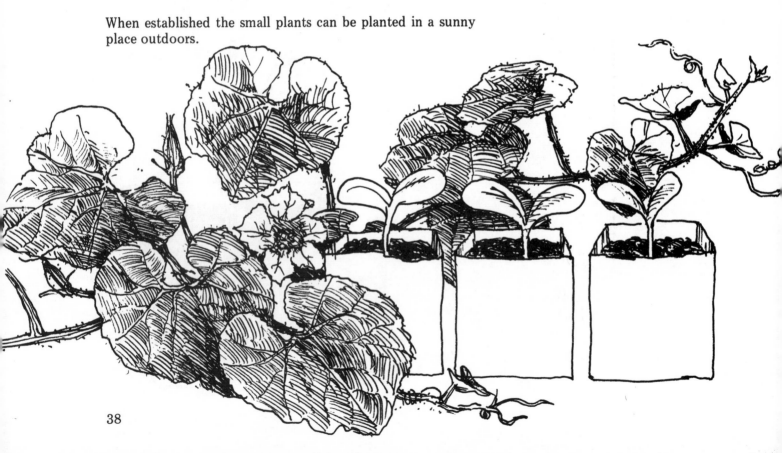

Sweet corn

Collect a ripe kernel of sweet corn. Allow it to dry out and then break the individual seeds from the corn cob.

Plant the seeds, either on cottonwool or in soil. When they have germinated, the children can transplant them into the ground.

If the corn is to be kept in the original container, the children could measure its growth rate from day to day.

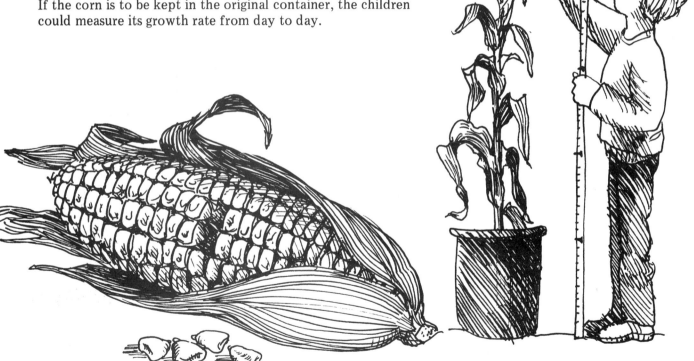

Oranges and lemons

The pips or seeds from oranges and lemons can be grown quite easily.

Repotting

Tap the pot to free the plant

- ☐ Soak the seeds in cold water for several days before planting them.
- ☐ Plant the seeds in light soil.
- ☐ Water daily, but do not let the soil get too wet.
- ☐ Keep the container in a light, sunny place.
- ☐ The seeds should germinate in about ten days.

The seedlings will grow too big for a small container.

Watercress

This interesting and nourishing plant can be grown from seed using a shallow tray or dish lined with blotting paper.

The paper must be kept moist and the tray should be drained and given fresh water each day.

The cress will germinate very rapidly in a sunny place, and a continual supply for eating can be maintained by planting fresh seed every few days.

Plants and people

Discuss with the children the belief which some people have that plants are sensitive to human contact.

Make one plant the subject of special attention for a set period of time. This might involve playing music to it, gently stroking its leaves, talking to it. Another plant, of the same age, type and size, would not get such special treatment.

After several weeks ask the children to examine each plant, consider its condition and give their opinions as to whether or not the treatment has affected the plant.

The plastic bag glasshouse

A large plastic bag can be used to cover any pot or plant. Because it shields the plant from draughts and increases the air temperature around the plant, this aids plant growth.

Seeds will also germinate quicker if shielded in this way.

In a simple way, this creates conditions similar to those found under glass in the terrarium.

Hanging gardens

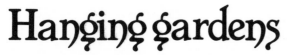

Many plants can be grown in small pots or containers which may be hung about the room or on a sheltered porch or even in the garden in some shady place.

Most plants will grow best in a large container which allows more room for root growth and also retains moisture longer. Any kind of pot, container or basket will do as a plant container. A light soil, well manured is best and care should be taken to avoid allowing the soil to dry out.

Ferns, begonias and most creepers and trailing plants do very well in such hanging gardens, which can be moved at different times of the year to allow the plants in them to catch more sunlight in winter and escape the fierce sun in summer.

Terrariums and bottle gardens

Bottles and a variety of glass containers make very good subjects for gardening in miniature.

Begin the project with a discussion about the possible materials which could be used in making bottle gardens. The origin of the bottle garden makes an interesting story which children will enjoy hearing.

Originally, the idea arose in Britain in the nineteenth century when botanists tried to find ways of sending plants home to England from remote parts of the globe. A man named Ward designed glass cases, which became known as 'Wardian Cases'. In this way rare plants and other botanic specimens could be carried great distances without damage as the sealed atmosphere of the case protected the plants inside from sudden changes of temperature and extremes of heat, cold, dryness or humidity.

This is the basic principle involved in any form of bottle garden, a sealed atmosphere which retains moisture and protects the plants inside from wind, heat, frost and the drying process. Any large, sealed glass container can be used to achieve the same result.

Making a terrarium

Although any glass container will do, it is desirable to use one which has a flat base, and is reasonably large. Generally, the soil used should be light and sandy.

Line the bottom of the terrarium with small stones or gravel, cover with sand, and add a light, fine soil. It is useful to mix a small quantity of charcoal with the soil as this helps eliminate souring acids.

Never over-water a terrarium as the water will stagnate and gradually poison the plants.

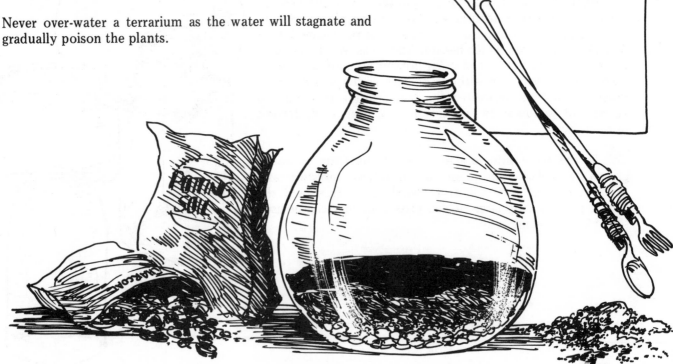

If the mouth of the bottle is small, improvise for tools

The terrarium

It is well to remember with the terrarium that this activity is one which can involve, as does all gardening, a combination of manual skills, scientific understanding and aesthetic appreciation. Children should be encouraged to understand the importance of creating miniature landscapes and helped to see the necessity for balance and perspective in the miniature world.

The terrarium can be of two kinds:

A desert terrarium

This is perhaps the easiest to establish because it does not have to be watered, except on rare occasions when a light watering or mist spray will be sufficient to keep the plants alive. In a desert terrarium only cacti or related succulent plants are grown. Sand and gravel help to give a suitably desert-like appearance.

The woodland terrarium

As their experience of bottle gardens increases, the children can set about establishing a woodland terrarium. This requires soil richer in nutrients than the desert terrarium. Any small ornamental plants usually grown indoors are suitable. Small ferns, African violets, mosses of any kind, or the smaller begonias all do well in terrariums. Bark, twig or rock arrangements are used to give a more realistic, woodland appearance.

Bonsai

Originally derived from China and Japan, bonsai gardening is the art of growing certain plants in small containers. The best subjects with which to start are members of the conifer family. Maples, palms, and some dwarf bamboo also give pleasing effects.

Before putting the plants into their permanent containers, usually small ornamental pots, their roots should be pruned so that they are confined to a small tuft around the base of the plant. The plant should then be potted into its permanent container in a normal soil. The container should be adequately drained and never over-watered.

A small landscape can be created around the bonsai plant using stones, mosses or twigs. Like the terrarium, the bonsai garden can be an exercise in landscape aesthetics.

Branches may be trained by wiring them into position.

Plant cases

An extension of the bottle garden is the plant case. Any glassed-in box can be used for this purpose. An ideal container can be made from any disused fish tank. Plants can be grown in the plant case in exactly the same way as in the bottle garden terrarium.

If the plant case is large enough it can be lined with sand and small potted plants, such as cacti, ferns or begonias, can be set in. If kept moist, there will be sufficient moisture in the sand to sustain the plants.

The advantage of this arrangement is that displays can be changed as different plants become available.

Window boxes

A wide range of plants may be grown outdoors or on window ledges if use is made of window boxes. These can either hold established pot plants or, if they are well drained, can be filled with soil and used as miniature gardens.

Generally it is quite easy to grow plants from seeds planted into window boxes, although sometimes it is easier to grow seedlings in a more sheltered place, and transplant them later into the window boxes.

Growing bulbs

One can grow bulbs in a plastic drink container, rather as people grow herbs in a 'florentine jar'.

Take a large plastic container, of the kind used for fruit juice, fill it with soil then cut several holes in the side and neck of the container. Plant a small bulb in each hole. Pierce the bottom for drainage and keep the container moist.

Herbs

Herbs are plants which have a special culinary or medicinal use. Most herbs grow well in pots or containers although they usually grow better in an open, sunny place rather than as indoor plants.

A sheltered corner, ledge, or window box are all suitable places for growing herbs. Some herbs are hardier than others and are better grown initially because they are both rapid and prolific in growth.

Amongst the easiest and best known is mint. There are a number of members in this family and all are hardy and vigorous. Other herbs which are easily grown in containers include thyme, sage, shallots, rosemary, marjoram and lavender.

Discuss with the children the various reasons for growing herbs and their uses; either for the flavour they give to food, drinks or sauces, or their fragrance which is added to soaps, oils and perfumes.

An interesting project for children would be research into herbs; their historical background, recipes—both ancient and modern, and their uses.

Mint — Moroccans like very sweet mint tea.

Thyme

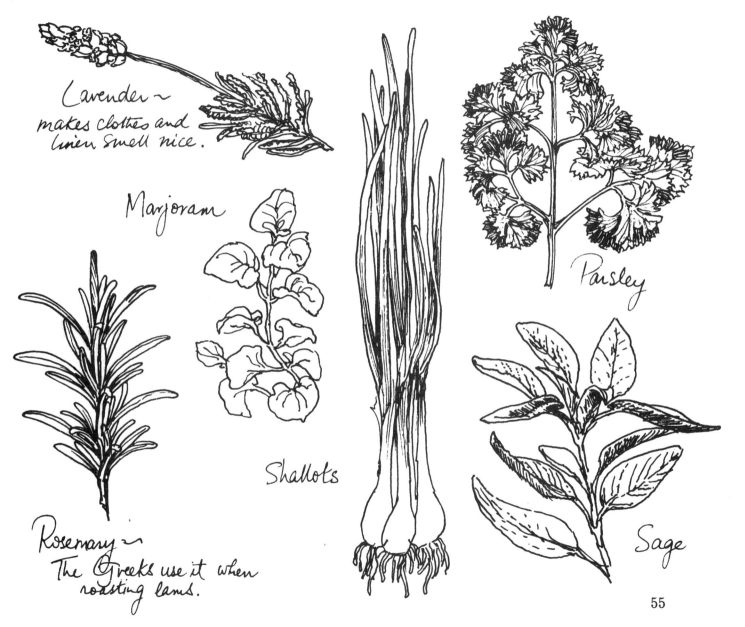

Lavender ~
makes clothes and
linen smell nice.

Marjoram

Rosemary ~
The Greeks use it when
roasting lamb.

Shallots

Parsley

Sage

55